The Jewish Colonies of South Jersey

Historical Sketch of Their Establishment and Growth

The Jewish Colonies of South Jersey

Historical Sketch of Their Establishment and Growth

&

Prepared and Issued by the Bureau of Statistics of New Jersey

William Stainsby, Chief

Camden, N. J.
S. Chew & Sons, Front and Market Sts.
1901

Republished for
The Alliance Heritage Center
by the South Jersey Culture & History Center Regional Press
Stockton University
2019

Republished 2019 for The Alliance Heritage Center
by The South Jersey Culture & History Center Regional Press
Stockton University

Stockton University
101 Vera King Farris Dr., Galloway, NJ, 08205

stockton.edu/alliance-heritage/
stockton.edu/SJCHC/

Copyright © 2019

ISBN-13: 978-1-947889-94-1

FOREWORD.

William Stainsby's report on the Jewish farming colonies in southern New Jersey, published in 1901 by the Bureau of Statistics of New Jersey, briefly describes the colonies of Alliance, Rosenhayn and Carmel nearly two decades after their founding; the report describes the Woodbine colony at greater length, ten years after its establishment.

The sketch of Alliance is reasonably detailed. Stainsby relates the colonists' early difficulties, mentions aid provided by the Hebrew Emigrant Aid Society and its successor the Alliance Land Trust, as well as the assistance of the Leach brothers of Vineland. The portrait of the colonists is compassionate and supportive. Stainsby praises both their perseverance and agricultural skills.

Rosenhayn, which was founded by six Jewish families in 1882, shortly after Alliance, had grown by 1901 to a village of 800 souls, with employment equally divided between industry and agriculture. The various manufactories paid their workers in cash, weekly; this was no company town paying wages in scrip. Stainsby describes the "untiring industry" of Rosenhayn's farmers and their "reward in ownership of the fine farms and the feeling of independence that emancipation from oppression and poverty brings."

The Woodbine Land Improvement Company, the farming and manufactory colony established in 1891 through the financial support of the Baron de Hirsch Fund, is the chief focus of the report. Stainsby is especially interested in the agricultural and industrial school, its buildings, dairy, poultry yards, and its course of study. He provides commentary on each of the manufactories as well. All photographs in this well-illustrated volume, except one, record aspects of the Woodbine colony. Stainsby clearly admires the measures set forth to instruct and train the rising generation "in theoretical and practical farming." Farming conditions in New Jersey differed greatly from those in eastern Europe. If "little could be done to bring the heads of families to understand and adapt themselves to these new means and methods, the new generation could be readily trained and instructed." This section of the report reads like a manual of best practices, c. 1900.

Stainsby describes the Jewish colony of Carmel as standing apart, in many respects, from Alliance, Rosenhayn and Woodbine. It was not initially funded by New York or Philadelphia philanthropists nor with support from the Baron de Hirsch. Farms were bought individually and financed through loans from the Building Association of Bridgeton. After several difficult years, the outlook for the colony was dire as that association began to foreclose on farm mortgages. If not for timely support from the Baron de Hirsch, Carmel would have failed. At the time of this report, though not wealthy, Carmel was on a sound footing.

Finally, Stainsby describes several failed colonies stretching across South Jersey from Alliance to Woodbine. Speculators initiated colonization schemes with limited capital, hoping to attract immigrants, pay low wages in various manufactories, and make quick profits. One after another, these communities failed to thrive. Stainsby provides their names, with spare details. Today they are intriguing footnotes to the Jewish colonies that did survive and, for a time, thrived: Alliance, Rosenhayn, Carmel and Woodbine.

William Stainsby, born in England on July 3, 1829, arrived in America at the age of two. A long-time resident of Newark, New Jersey, he trained as a hatter, spent several years in the saddlery and hardware business, and eventually became engaged in the sale of oils and paints. An ardent Republican who joined the party in 1856, he enjoyed a lengthy political career, serving as alderman in Newark from 1866 to 1874 and again from 1890 to 1894. He represented Essex county in the state senate from 1882 to 1884. In 1898 he was appointed Chief of the State Bureau of Labor and Statistics for a five year term. Under Stainsby's management the bureau became a prolific branch of state government, publishing numerous useful studies.

Stainsby's report provides a contemporary view of the earliest Jewish farming settlements in South Jersey; the sketch is brief but illuminating. It does not, however, convey the rich complexity of these early colonists' experiences. For that, readers are directed to the section on Further Reading at the conclusion of this volume.

This slim volume, the first published by the Alliance Heritage Center at Stockton University, is meant to serve as a companion to *Migdal Zophim & Farming in the Jewish Colonies of South Jersey*, published in 2019. These titles are the first offerings in a series of works exploring the history of Jewish farming in South Jersey.

CONTENTS.

The Jewish Colony at Alliance 9

Rosenhayn 13

Woodbine 17
    The Baron de Hirsch Agricultural and Industrial School 18
    Buildings 20
    The Dairy 21
    Poultry 22
    Three Years' Course 23
    The Faculty 23
    The New College 24
    The Woodbine Land Improvement Company 24

Carmel 29

Other Colonies 33

Illustrations 8, 16, 35–49

Further Reading 51

Brotmanville factory encompassed by the tailoring settlement.

The Jewish Colonies of South Jersey.
Historical Sketch of Their Establishment and Growth.

THE JEWISH COLONY AT ALLIANCE.

The first Jewish colony in South Jersey was located in Pittsgrove township, Salem county, in 1882. When the persecuted Jews were driven from Russia a number of wealthy and influential Hebrews in the city of New York formed the Hebrew Aid Society, whose object was to give assistance to their fellow religionists who were thus cast out from the places of their birth and compelled to seek shelter and subsistence among strangers in a strange land.

This society purchased about eleven hundred acres of land in Pittsgrove township, in the county of Salem, a little over six miles northwest of the borough of Vineland, Cumberland county, and about two miles from Bradways Station, on the line of the New Jersey Southern Railroad. The society contracted with Messrs. George Leach & Bro., Vineland, to erect three temporary buildings, or barracks, each to be 24 feet wide and 150 feet in length. Only three days were given for the erection of these buildings, which were of the rudest possible character, but longer time could not be granted, as the emigrants were on the way and shelter must be ready for them upon their arrival. In a few days the exiles landed from the steamer and were conveyed to the site of the new colony.

They had been cast out as paupers; their humble homes in Russia had been taken from them, and they fled as did the Pilgrim fathers from tyranny and relentless persecution to a land they knew not, but with the promise of such assistance as would enable them to make homes for themselves and children, and where they would be free to worship God in their own way, assured of liberty and the protection of the laws. The expense of their transportation was defrayed by the society and they came with the most scanty supplies of clothing and food and took shelter in the rude barracks which had been so hastily erected for their accommodation.

The refugees numbered about two hundred and fifty men, women and children and they marched stolidly along over the field from the railroad to the location where they were to carve out for themselves new homes of greater comfort than they had ever before been able to enjoy. They were very secretive people and it was found that, notwithstanding they had been robbed, outraged and abused by the Russian officials, some of them had managed to hold on to a little money, which was very helpful to them in the new land. The exiles excited a great amount of curiosity among the people of the vicinity by their humility; they would doff their hats on the approach of a neighboring farmer or visitor and stand with bowed heads, as if they feared every moment to feel the blow of the knout or hear the harsh voices of the Russian officers; that, however, is a thing of the past; they have lost their servile appearance, but are still quite courteous and polite to visitors.

The society proceeded to allot the land in tracts of fifteen acres to each family, on which before the winter set in, humble cabins, twelve by fourteen feet, were built and occupied by the families; in the case of a large family a lean-to was added. The society deeded these farms to occupants charging each one hundred and fifty dollars, and giving the term of payment at thirty-three years, without interest.

After two or three years the Hebrew Emigrant Aid Society gave place to the Alliance Land Trust, which gave its name to the settlement. Not only did these prosperous Hebrew merchants of New York thus provide small farms for these outcast people, but they sent a committee from New York to Vineland to ascertain if anything further was necessary to be done for them; singular to say not a single member of the colony came to Vineland to meet the gentlemen from New York. After a lengthy interview with Mr. George Leach, of Vineland, who was very thoroughly acquainted with the condition of the colony, the committee returned to New York, and the result of their visit was the immediate remittance of Mr. Leach of the sum of $7,000, to be distributed, $100 to each of the seventy families, to enable them to secure the agricultural implements needed.

Of course, it was necessary to give support to these people through the first winter. The open days were utilized by them in clearing the land, freeing it from stumps and getting it ready for cultivation. Through the winter some work was done in the houses in making clothing, to which these people seem most readily to adapt themselves; manufacture of cigars, knitting of woolen caps, capes, etc., by which the families were enabled to earn something toward their support. In the spring, active farming and trucking operations began, and from that day to the present time the result has been a steady uplift and improvement in the moral, social and financial condition of the people. Can a Jew become a successful farmer, is a question frequently asked and almost invariably answered in the negative, but a careful and impartial investigation of the work accomplished by these colonies will justify a more hopeful conclusion. A visitor will observe good houses, improved and thoroughly up to date outbuildings, healthy and well-conditioned stock, and crops growing that are admirably adapted to the character of the soil. These and other details of management open to observation, which show a high degree of intelligently directed industry, will justify the assertion that the Jew not only can, but has become, a successful farmer, at least in these settlements.

The soil at Alliance is a light, sandy loam, not well adapted to cereals, of which but little is raised except a small quantity of corn for home use, but it is as good as any in the country for growing fruits, berries, grapes and sweet potatoes, and to these from the very beginning the people of Alliance have turned their attention with marked success. They raise very fine strawberries, raspberries, blackberries, cherries, pears, peaches and immense quantities of sweet potatoes of very excellent quality. The main market for their berries and fruits is New York, shipments being made by the New Jersey Southern Railroad, which has made careful arrangements for the prompt forwarding and delivery of consignment, thus enabling the farmers to get their produce to market in good condition and consequently get fair prices.

The convenience of shipment at Norma and Bradways' Station, which are so located that the mass of farmers do not have to drive over two miles, is a great advantage, as the berries and fruit are not injured by being carried long distances over rough roads, and reach New York markets sound and fresh. The sweet potatoes raised at Alliance have attained such high repute in New York that they command from twenty-five to fifty cents per barrel more than can be obtained from those raised elsewhere.

The farms have a very neat appearance and give evidence of great care in cultivation, no rubbish being permitted to accumulate. The vineyards have been carefully laid out, the vines are healthy and strong and the yield is very large; but little attention is given to wine making, as shipments of the grapes in fresh and sound condition to New York markets is found to yield more satisfactory results.

In the shipment of strawberries, raspberries, blackberries, etc., great care is exercised in selecting and packing, and they have thus secured a good reputation in the markets.

The farmers of Alliance have good stock, the cows especially being of the very best; the poultry also will compare favorably with any in this section of the State. As cows and poultry are prime factors in solving the problem of family subsistence, they receive a vast amount of care and attention. The Jew farmer will give the stock the best to be obtained and the strictest attention to its comforts and health to the verge of his own self denial. Special details of items of crops could not be obtained, but the berry and fruit crop of 1899 amounted to $40,000. The sweet potato crop realized for these thrifty farmers $18,000.

Manufacturing in Alliance has not advanced as rapidly as in the later colony at Woodbine; there is one large factory, which is operated by the Alliance Cloak and Suit Company, of which Mr. Abraham Brotman, a thoroughly wide awake and progressive man, is the head. The factory is located on the northern portion of the tract, which is known as Brotmanville.

In this factory, A. Brotman employs fifty-five hands; T. Eskin thirty, and T. Brod fifteen. They are all engaged in the manufacture of ladies' and children's coats and cloaks. The operatives average about $12 per week, and the wages are paid weekly and in cash. A large new three-story factory has been erected a short distance from Mr. Brotman's, but is not yet occupied.

The colony at Alliance has had a hard struggle, but has passed the experimental stage and is now fairly on the road to success. It has recently passed from the control of the Alliance Land Trust to the Board of Trustees of the Baron de Hirsch Fund; these trustees propose to extend immediately material aid to the colony. They will spend $10,000 in public improvements and build twenty fine dwellings. This, the first colony establishment in South Jersey, has not had the success which has crowned the colony at Woodbine, but it must be remembered that Alliance has not had, hitherto, the benefit of large appropriations from the Baron de Hirsch Fund as have been given to the people of Woodbine.

ROSENHAYN.

In 1882 the land now occupied by the prosperous Jewish colony of Rosenhayn was a wilderness of pine and bushlands. The Hebrew Emigrant Aid Society of New York City, which established the colony of refugees at Alliance, located six Jewish families at this point, which has now grown to be a village of some note with a population of 800. It is located on the New Jersey Central Railroad, midway between the cities of Bridgeton and Vineland, and about two and a half miles northwest of the village of Carmel. The town site has a broad, well-shaded avenue over one mile in length, with excellent sidewalks and very few cross streets; this avenue runs directly from north to south and fronting it are nearly all the dwellings.

Rosenhayn has a more elevated position than Carmel; there are no swamps, and with proper sanitary regulations the settlement should be in a healthy condition. It has railroads, express, telegraph and telephone offices; there are some two hundred houses, with about two hundred and thirty families resident in the village and on surrounding farms. The population is composed almost exclusively of Russian and Polish Jews, who, freed from the oppression and tyranny to which they had been so long subjected, are rapidly advancing in intelligence and acquiring a higher degree of civilization. The population of Rosenhayn is about equally divided between industrial and agricultural pursuits; there are nine manufactories; the articles manufactured are clothing, hosiery, foundry work, tinware and brick. The number of hands employed in these factories is divided as follows: clothing, 160; brickyard, 17; hosiery, 5; foundry, 4; tinware, 2. The average wages of the operatives is $10 per week, which is paid every week in cash; there are no stores connected with these factories. The character and condition of the dwellings of the workmen is good; about fifty per cent. of the employees own the houses they occupy.

The proximity of Rosenhayn to New York and Philadelphia insures these colonies a large amount of work in the fall and winter seasons from these cities, but at very inadequate wages. The work made in the clothing factories is principally shirts, ladies' wrappers, cloaks and white goods of various kinds. Careful attention is not paid to ventilation, and when the condition of the weather requires the closing of the windows, the air in the shops is very impure. In addition to the factories, garments are made in many of the homes.

The farming portion of the community appears to be fairly prosperous. Of the 1900 acres comprising the tract, about one fourth is under cultivation; the farms are in excellent order and exhibit evidences of skillful manipulation in clearing the soil of stumps, roots and noxious weeds. The soil, as in other colonies, is not well adapted to the raising of cereals, and the attention of the farmers is given to the culture of fruits and vegetables. The shipment of berries, sweet and white potatoes and other vegetables to the New York market is very large, and the railroad station presents an animated scene as the farmers bring in their produce on shipping days; large quantities of grapes are also raised for shipment; wine making is largely carried on, and the vineyards, being carefully cultivated, present a thrifty and strong appearance. The great source of profit, however, is the sweet potato crop; the yield is enormous and of such fine quality as to command the very highest prices in the New York market. The farmers here are planning for the construction of a canning factory to avoid the shipment of berries and tomatoes. The farmers of Rosenhayn are hard workers and do not count the hours of labor; from the earliest dawn until sundown they are hard at it, and their untiring industry is winning its reward in ownership of the fine farms and the feeling of independence that emancipation from oppression and poverty brings.

About fifty per cent. of the farmers have their farms clear of incumbrance; it was a hard struggle and uphill work for years, but their perseverance and economy have at last brought them to a fair degree of success.

The farmers of Rosenhayn have good stock and keep it in excellent condition; a Jew may be trusted to take the best possible care of his horses and cows; he regards them as very potent factors in winning his way upwards, and they are treated as well as the family. Considerable attention is paid to poultry raising, and as in the case of the other colonists, these people seem to have the knack of doing it well. The farm dwellings are small, but with their surroundings are neatly kept, and the outbuildings are also in reasonably fair condition. The annual value of crops raised is between $10,000 and $12,000. There is now no question but that the Jews can make a success of farming. These colonies located in South Jersey have demonstrated that fact beyond controversy.

It must be remembered that these people came here in the condition of paupers with but little experience in farming and that little acquired under entirely different circumstances of climate, soil and farming methods, but they have proved to be apt pupils. The very liberal aid extended to the colonists in starting them out is a great incentive to industry, economy and perseverance.

The land is at first divided into small farms, small buildings erected and a family is given one rent free for two years; after that small payments are required annually until the farm is paid for in full.

Jewish Colonists at Woodbine.

THE JEWISH COLONY OF WOODBINE, CAPE MAY COUNTY.

In "The Study of the Jewish Colonies of South Jersey," there is none that presents features of more absorbing interest than the settlement at Woodbine which is the direct application, in practical form, of the philanthropic and beneficient designs of the late Baron de Hirsch, which have been faithfully considered and carefully carried out by the Board of Trustees of the great fund which the late Baron bequeathed for the amelioration of the condition of the persecuted Jews of Russia and their uplift in business, moral and social life; it bears the title of the Woodbine Land and Improvement Company.

The experiment of planting a colony in the bushland of South Jersey was certainly a bold one; to take the bush and woodlands, clear and subdue the soil and work it to productiveness and fruitfulness was a stupendous enterprise and required careful thought and planning, ceaseless and untiring activity and energy to produce satisfactory returns. It has been, however, as the facts detailed in the paper here presented will establish beyond cavil or dispute. This settlement was mapped out in 1891.

Woodbine is located in Upper Township in the northwestern section of Cape May county; it is fifty-six miles from Philadelphia and twenty-five from Ocean City and Atlantic City. Two railroads—the West Jersey and Seashore and the South Jersey—give direct communication with the neighboring towns and with Philadelphia and New York. The tract comprises 5,300 acres, 1,800 being now cleared and improved; the soil is of a loamy character, being a mixture of sand, clay and gravel, suitable for such crops as fruits, vegetables, rye, oats, clover, grass, etc. The soil, naturally warm and level, is easily worked and when once cleared of stumps and roots is subject to easy drainage and well manures and fertilizes; it is rich in the mineral constituents though somewhat deficient in humus owing to the frequent forest fires, but this being remedied by a resort to green manuring, i.e., the plowing under of crimson clover, rye, buckwheat, etc., the form of which is here of very full and luxuriant growth: when sown in the middle of September it is ready to cut early in the following May thus giving opportunity for a second crop of sweet corn or potatoes. There is no haphazard farming on this tract, but all conditions of the soil have been studied and the operations shaped accordingly. Analysis of all manures and fertilizers has been made and only those are used which exhibit the greatest adaptability to the nature of the soil and which it most readily absorbs.

THE BARON DE HIRSCH AGRICULTURAL AND INDUSTRIAL SCHOOL.

The trustees of the fund recognized the fact that measures must be taken for the instruction of the rising generation and their training in theoretical and practical farming; the conditions existing in this country were entirely different from those of the land from which they came and while little could be done to bring the heads of families to understand and adapt themselves to these new means and methods, the new generation could be readily trained and instructed.

The organization of this school was effected in October, 1894, after a considerable preparatory work had been done by a class made up of the sons of settlers located upon the tract, who spent the fall and winter of 1893 and the spring and summer of 1894 in clearing and improving the land.

These boys were also given instruction in English, arithmetic and other subjects; during the winter months a series of lectures was given on various practical agricultural subjects, illustrated by stereopticon views, once in each week. These lectures the parents of the boys were permitted to attend.

During the preparatory period of the year (March to October, 1894) forty-two students were registered. The erection of the buildings, all excepting the large school, was mainly the work of the future students. In October, 1894, the first regular class was organized; the course of instruction combined theoretical teaching and practical application; the boys were taught the English language, history of the United States, arithmetic, drawing, land measuring, botany in its application to horticulture, chemistry in its relation to soils and crops, the proper feeding of domestic animals and entomology. The practical portion of the work was carried on mainly in the greenhouse, and included the preparation of soils, the potting of flowers, propagation by seeds and cuttings, preparation and care of cold frames and hot-beds and the grafting of roots.

The land was in good condition in the spring of 1895, and the fifteen students had practical application of their winter studies in getting in the seeds and plants, which comprised fruit trees, grape vines, strawberries, raspberries, blackberries, currants, gooseberries, onions, cabbage, potatoes (white and sweet), carrots, beets, peas, beans, sweet corn, in fact vegetables of all kinds, together with broom corn, millet, sugar beets and several kinds of grasses, which were grown for commercial and experimental purposes. During the four months, from October, 1894, to February, 1895, the students spent from twelve to fifteen hours per week in the machine shops, where they became acquainted with the handling of tools and the operating of machinery.

The second year opened with a class of twenty-two. The buying and mixing of manures and fertilizers, with instructions in comparative anatomy and physiology, were added to the curriculum. Great care was bestowed upon the orchards and vineyards and the growing of forage plants, made necessary by the absence of meadows and natural pasture lands. The end of the second year brought with it assurance of the success of the methods employed. In September, 1896, an exhibit was made at the Cape May County Fair of corn, preserved fruits, peaches, grapes, melons, flowers, floral designs, poultry, etc., representing the average of the farm products. First premiums were awarded the school for all the exhibits except corn, and for this second premium was given. This year there were twelve graduates: one entered Storr's Agricultural College, Conn.; two were retained as assistants on the school farm; one secured a position as gardener at the Jewish Hospital in Philadelphia; another as florist in New York; the *remainder went to work on the farms of their parents*; certainly a very promising result.

The third year opened with twenty-one students; the course of studies was enlarged and the hours of practical work increased. The acreage under cultivation was as follows:

Orchards . . . . .13 acres.
Blackberries and raspberries. . . . . . 6¾ acres.
Strawberries . . . . .6¼ acres.
Corn and forage plants . . . . 21 acres.
Vegetables. . . . . 6 acres.
Grapes . . . . 4 acres.
Nursery . . . . .1 acres.

Three acres of strawberries in bearing yielded an average of 3,800 quarts per acre; the best acre gave a total of 4,728 quarts.

The fourth year opened in October, 1897, with twenty-one students; the studies were about the same as those of the previous year, but an addition was made to the practical work in the care and management of the dairy, testing of milk by various methods, poultry raising, etc., instruction in higher mathematics. During the winter two exhibits were made—one at the Annual Fair of the Hebrew Literature Society of Philadelphia, where several diplomas were awarded, the other at the Washington Feather Club, of Washington, N. J., where the school received first and second premiums for poultry. At the close of the year several of the graduates secured excellent positions as instructors.

There are now in the school ninety-six students, several of them girls; the latter, in addition to their educational studies, are being instructed in housekeeping, dairying and floriculture. The school farm contains 270 acres, 175 of which are under cultivation, and, with the growing crops, presents a very attractive appearance. The yield of the farm in 1899 was as follows:

White potatoes . . 110 to 175 bushels per acre.
Sweet potatoes . . . . 35 to 50 barrels per acre.
Corn . . . . 25 bushels (shelled) per acre.
Tomatoes (partial failure) . . . . 3 tons per acre.
Peas, beans, onions, salad, etc., very large yield.

FRUITS.

Strawberries . . . . avg. 2,500 quarts per acre.
Raspberries . . . . avg. 1,600 quarts per acre.
Blackberries . . . . avg. 3,000 quarts per acre.
Grapes . . . . .avg. 3,000 pounds per acre.

The orchards are quite young, but the average showed:

Peaches . . . . 4 baskets per tree.
Pears . . . . 5 baskets per tree.
Plums . . . . 6 baskets per tree.

Apples were not yet bearing. The promise for 1900 of all these fruits is for a very large yield, as all the trees are well set.

BUILDINGS.

The buildings on the farm have been carefully designed for their respective purposes; the offices are located in a large frame building, with four rooms on the lower floor and school room, library, chemical cabinet, library and reading room on second floor. The library is well supplied with books on science, agriculture, history, reference and general literature, with numerous maps, charts, etc.

The reading room has a full supply of daily and weekly papers, magazines, etc. The chemical laboratory is where the constituent elements of products are determined and arranged in cases and analysis of fertilizers are made, and the results preserved in jars.

Hirsch Hall, the home of the students, is a building 46 x 72 feet, three stories high, with large verandas, and is admirably arranged and fitted; the basement has a large dining-room for the boys, lavatory laundry, supply, storage rooms, etc.; the second floor has large reception room, parlors, teachers' dining-room, and two large dormitories at the extremes of the building facing respectively north and south. These are very neat and clean, fitted with iron cots with wire springs and hair mattresses. Rooms for the matron and teachers are also on this floor. The third floor has store rooms and two dormitories, the same size and immediately over those on the second floor, and are fitted in similar manner. Careful attention is paid to sanitary conditions, light and ventilation. The dormitories, being constructed at the extreme ends of the building, with large windows and air shafts, receive light and air from three sides. There are accommodations for twenty-eight students in each of the dormitories; the heat is supplied from the boiler house, which is located in a separate building; the water supply is ample and large pipes extend up to third floor, where also is fire hose on reels ready for instant use. Fire escapes are also conveniently arranged for the exit of the inmates in case of fire, making the building in all points thoroughly equipped.

THE DAIRY.

The dairy provokes the admiration of all who visit it, perhaps there is none in the State where the sanitary conditions are more carefully looked after. The building is 30 x 60 feet, provided with every modern appliance for the care and comfort of the stock.

At the southwest corner of this building the students have constructed a silo thirteen feet in diameter and twenty-nine feet high, with a capacity of eighty tons. In the construction of all the buildings on the farm, the students have done a large part of the work.

The front of the dairy building is two stories high, containing feed room, mows and large creamery, with all the latest improved machinery for the treatment of the milk and making butter. Here the milk is brought direct from the cows, placed in a cooler which reduces the temperature to fifty degrees or less, sterilized, bottled and placed in the refrigerator ready for sale, or goes to the separator. The butter made here is of a very superior quality, its purity being guaranteed, and commands the highest prices. The creamery contains all the machinery, operated by steam power, for treatment of the milk and making butter. A guaranteed analysis of the milk is furnished to customers semi-monthly. The herd consists of twenty cows, six heifers and two bulls; the cows are mostly Alderneys and Jerseys; one bull is a registered Guernsey, the other a Holstein. The quantity of milk given by each cow at milking is weighed and noted on the register; frequent analysis determines the quality of the milk of each cow, a portion being daily retained for that purpose. Every precaution is taken through all the operations to secure the absolute purity of the milk. The students assigned to do the milking, when that time arrives, are required to give their hands and arms a thorough washing, clothe themselves in white canvas suits and then proceed to work.

No one, not especially designated, can enter any part of the dairy or go through it in passing from one building to another. The building is completely screened with wire netting, so that the animals experience no annoyance from flies, gnats or mosquitoes. A multitude of windows make the cow house very light and airy; steam pipes, running through the building, maintain a comfortable temperature in the winter. The average amount of milk daily from each cow is about two gallons.

The cows are very carefully groomed twice daily; the cow house is kept conspicuously neat and clean; there is a trough running through the heads of the stalls the entire length of the building to give a constant supply of fresh running water to the stock; the stalls are fitted with patent yokes, fine feed boxes, salt cups, etc.; a trench back of the cows carries off the voidings of the animals to manure pits outside the house. The herd is carefully inspected once every month by Dr. Tremaine, of Bridgeton, and if any cow is found ailing in any way it is at once isolated from the herd and so kept until it recovers or is condemned. A wagon conveys butter, milk and vegetables to Ocean City through the season. The milk from two or more cows is set apart for the use of infants.

POULTRY.

The school has scored a great success in the raising of poultry. There are four large poultry houses, and four more being constructed. These are all models in arrangements and appointments; incubators of the latest and very best models are in use, and stretching out from these are the wire screened apartments wherein the different broods are confined; the age of any chicken is thus determined by the number of the apartment he occupies.

Enclosed steam pipes give the poultry house a moderate temperature in the coldest weather. The sanitary condition of these houses is perfect; they are thoroughly cleaned every day by the students who have that turn of duty. The poultry raised here is pretty sure to command a premium where ever exhibited.

It would make this study too lengthy to speak in detail of the large greenhouses for the growing of flowers and vegetables, the barns or the new creamery building, which is now being constructed, as well as a large barn for the accommodation of the eight horses employed on the farm. There are also blacksmith and wheelwright shops; in the former the farm tools are made, rehandled and repaired; in the latter the wagons to be used on the farm are made and repaired.

THREE YEARS' COURSE.

A brief synopsis of the students' three years course is of interest:

First Year—Winter Term—School twenty-eight hours per week. Taking care of stables, poultry yards and domestic animals; milking; work in wood working shop and in the fields, twenty-eight hours per week.

Summer Term—School fourteen hours per week. Care of stables and animals; work in field and garden; planting and taking care of the growing crops and harvesting; eight hours per day, five days in week.

Second Year—Winter Term—School twenty-eight hours per week. Work in the greenhouses, cold frames, hot-beds; work in the orchards, including trimming and grafting; work in the blacksmith shop; twenty-eight hours per week.

Summer Term—School fourteen hours per week. Care of orchards and small fruit plantations, greenhouses and open ground floriculture and work on the nursery grounds; eight hours per day, five days in week.

Third Year—Winter Term—School twenty-five hours per week. Continuation of work of second year and work in the wheelwright shop; thirty hours per week.

Summer Term—Undergraduates are sent out to farms in New Jersey and other States for practice and to familiarize themselves with local conditions of farming. The demand for these boys is always in excess of the supply, and they readily command $20 per month and board.

The course for the girls comprises the same hours for study in the school, the remaining time is devoted to sewing, cooking for the boys, caring for the poultry, dairy work and instruction in housekeeping under direction of the matron.

THE FACULTY.

Prof. H. L. SABSOVICH, M.A.,
Superintendent and Instructor in Theoretical Agriculture.
THOMAS E. GRAVATT, B.S.,
Instructor in Mathematics and English.
JACOB KOTINSKY, B.S.,
Instructor in Natural Sciences.
JOSEPH PINCUS, B.A.,
Farm Superintendent and Instructor in Dairying.
FREDERICK SCMIDT,
Instructor in Horticulture and Floriculture.
SIMON BRAILOWSKY,
Instructor in Wood and Iron Work.
ACHILLES JAFFE,
Instructor in Religion.

There are now ninety-six students, several of the number being girls. Every applicant for admission must be at least fourteen years old, in good health, and must submit to the Superintendent testimonials of good moral character. Applicants must be prepared to pass the third grade examinations as given in the public schools of Cape May county. Tuition is free to all regular students; board and lodging at actual cost. To those students whose parents are unable to support them while at school, board and lodging will be given gratuitously and offset by the labor of the students on the school farms; students must furnish their own clothing. Voluminous reports are prepared by Prof. Sabsovich, showing every detail of the work in school and on farm, and are submitted to the Board of Trustees at the close of each term.

THE NEW COLLEGE.

Two years ago it was found that the school building was entirely too small to accommodate the rapidly increasing number of students, and the Board of Trustees of the Baron de Hirsch Fund was applied to for an appropriation for a new building; this was granted, and the Jewish Colonization Society of Paris also made a very liberal contribution to the building fund of the new college.

It is now being built and will be completed and dedicated with appropriate ceremony in September. The building is of brick, 66½ x 78 feet, and three stories high; the basement story is ten feet high and will contain gymnasium, bowling alley, bath rooms with shower baths, wash rooms and water closets, bicycle storage room and boiler room, etc. The second story is thirteen feet high and will have a fine hall, reception room, parlor, chemical laboratory, two large school rooms, photographic room and two cloak rooms. The third story is fifteen feet high with large hall, two large school rooms and an assembly room for lectures and entertainments, which will seat about three hundred persons. There is also a loft for storage, thirteen feet to the peak, and is surmounted by a fine belfry. The front entrance will be very ornate with large arched twin windows each side with stained glass. The very latest and best arrangements will be adopted for light, heat and ventilation, also drainage; the furnishings will be of the best.

The grounds will be made beautiful with trees, shrubbery, plants and flowers. The cost of the building will be about $25,000, exclusive of furnishings and apparatus; it is to be made thorough and complete in all its appointments. Hirsch Hall, which is near the new college, will still remain as the boarding house of the teachers and students.

THE WOODBINE LAND IMPROVEMENT COMPANY.

The incorporation of the Baron de Hirsch Trustees for this settlement bears the above title.

The 5,300 acres comprised in the tract were purchased on the 28th day of August, 1891; the first settlers came in the spring of 1892, and consisted of fifty families, in all about three hundred persons. The colonists were brought on from Northwestern and Southwestern Russia and Roumelia; to each family was assigned fifteen acres of land, with the privilege of acquiring fifteen acres more, if they desired to do so. It was originally designed for a purely agricultural colony, no manufactures being contemplated, but as the school and other farms became productive and the farmers sought to dispose of the surplus of the products above their family needs the fact was recognized that where a battalion of producers was created it was absolutely essential that there should also be a bridge or division of consumers. This condition of affairs was promptly seen and was immediately provided for.

The town site was laid out in 1897, comprising 800 acres, 275 of which have been cleared. Manufactories were located by erecting buildings, the corporation granting such concessions as induced manufacturers to remove to Woodbine. Houses were built for these operatives, and today the town contains a population of over fourteen hundred.

There are one hundred and sixty Jewish and thirty-four Gentile families; fifty per cent. of the people own their own homes. Forty per cent. of the population is engaged in agriculture and sixty per cent. in industrial pursuits. Of the business men, twenty in number, fourteen own the properties they occupy; of the farms, fifty in number, sixteen have been entirely cleared of incumbrance and deeds given; the remainder are under lease and the farmers are rapidly extinguishing their indebtedness, $1,100 having been paid off in the last four months.

The townspeople are taking great pride in the improvement of their properties, and are setting out shrubbery and ornamental plants and flowers.

The public buildings comprise a synagogue, erected at a cost of $7,000; a Baptist Church, cost of building, $2,500, for which the ground was generously donated by the Woodbine Company; a public bath-house, built at a cost of $2,500; two school buildings, one built by the colony at cost of $2,500, the other by the township at a cost of $2,000. There is a fine hotel opposite the West Jersey and Seashore Railroad Company station. The total valuation of property in Woodbine is $250,000, of which sixty-three and one-half per cent. is owned by the Woodbine Land Improvement Company and thirty-six and one half per cent. by other parties; the value of the farm property is $75,000.

The manufactories are as follows: The clothing factory of Messrs. David & Blumenthal, manufacturers of mens' and boys' clothing. The factory is 50 x 60 feet, three stories high, with a smaller one-story building 30 x 40 feet; they employ one hundred and sixty-eight hands, with a weekly payroll of $1,000. If they could get sufficient hands they would more than double the capacity of the factory. In a little over one year this firm has built thirty-four houses east of the railroad for their workpeople. Their method is one very favorable to the employees. When a workman wants to own a home of his own, he announces it to the firm and pays down $25; the house is then built. When it is ready for occupancy he makes a further payment of $50 and is permitted to move in, taking a lease at $7 per month rental, which is credited on the house as paid, so that when the payments amount to a sum sufficient to cancel the balance remaining on the property he is given a deed for it; no interest is charged. These houses are built at a cost of about $525.

The Universal Lock Company employs forty hands; weekly payroll of $360. The Woodbine Machine and Tool Company employs twenty-eight hands, with a weekly payroll of $225.

Louis Schuleman, manufacturer of clothing, employs eighteen hands; weekly payroll $108.

The Woodbine Brick Company employs twelve hands; weekly payroll of $100.

L. Rosine manufacturer of clothing, employs five hands and pays out weekly $40.

The average earnings of each family on the tract is a little over $500 per year. There are in the town nine carpenters, four bricklayers, five painters, twenty-four other mechanics. The percentage of the townpeople as to employment is as follows:

Clothing Factories .... 50 per cent.
Machine Shops .... 25 per cent.
Building Trades .... 12 per cent.
Storekeepers, Teachers, etc. .... 13 per cent.

There are fourteen miles of streets laid out, four-miles of which have been graded and gravelled; there are twelve miles of farm roads laid out, improved and in excellent condition; an electric light plant has been installed, the power for which is furnished by the Woodbine Machine and Tool Company, which also furnishes the power for the factories; lights are furnished for bath, public and private use; the streets are lighted by twenty arc lights. There are no running streams or surface springs on the Woodbine tract; the water supply is secured from artesian wells, the water from which is pumped into two large tanks, one containing 30,000 gallons and the other 18,000. When full these give an average of 34.29 gallons per capita daily for every man, woman or child on the tract and is excellent water.

The community is orderly and law respecting and a single policeman is considered a sufficient safeguard.

A very close and thorough inspection of every part of this colony, both as to its agricultural and industrial conditions, gives conclusive evidence of the industry, thrift and economy of the people.

The industries are somewhat hampered by the inability of the manufacturers to obtain a sufficient number of operatives to increase their output; the workmen apparently prefer the sweat shops of New York and other large cities with their noisome air, confined quarters and reeking filth of their surroundings to the commodious, well-lighted, thoroughly ventilated factories and the free air of the open country. One reason why it is so difficult to get these people away from the large cities lies in the fact that although the manufacturers have labored earnestly to induce them to come out into the country to work, it has been found impossible to divert their minds of the fear that the employment will not be permanent and they may be thrown out of employment without means and far from their friends and associates.

The Jews who enter upon farm life are hard workers, and from earliest dawn to sundown the hours are spent in the labor of the farm. They are always anxious to find the best methods to pursue in cultivation of the soil and the treatment of growing crops. In taking a tract of fifteen acres for his farm, the head of the family devotes himself to that work, perhaps retaining a son to help him, the rest of the children find employment in the factory and earn sufficient to supply the needs of the family until the farm is well cultivated and productive. There are few drones in the Jewish hive. One of the finest farms in Woodbine is that of Farmer Lipman, which adjoins the northwest corner of the school farm. It is on slightly rising ground, with a slope to the south, and has been brought to a very high degree of cultivation. His crops are very fine and every portion of the land shows the greatest care exercised and the result of the hard labor that has been expended upon it. Mr. Lipman's comfortable dwelling is surrounded on all four sides by a high grape trellis; the house thus standing, as it were, in a large court, with walls of greenery; the effect is very pleasant and attractive.

This farmer has one daughter who is clerk to the school and another earning fair wages in the large clothing factory. There are several other farms in almost equally excellent condition.

Yes, the Jew can be made a very successful cultivator of the soil; he bears the elements of success in his quickness to learn; his ready adaptability to the circumstances by which he is surrounded; his untiring energy and close economy. To assert the contrary is to betray the effects of prejudice and not conviction brought about by a knowledge of facts.

The question of erecting an elegant marble monument to Baron de Hirsch has been agitated at the Woodbine colony, but what need of that? Here, grander than marble shaft or column, more enduring than mausoleum of granite or polished tablet of brass is this thriving settlement, made up of a people rescued from tyranny and despotism, raised from abject poverty to become, by the aid of his beneficence, intelligent citizens of the Great Republic, industrious, enterprising, economical and self-respecting. It was a noble bequest; it is being faithfully applied to the purposes of the trust, and the promise is for the highest and most gratifying results. This is his monument.

THE JEWISH COLONY AT CARMEL.

The Jewish colony at Carmel is in many respects different from those at Alliance and Woodbine. There was no purchase of a large tract of land for division among those who came to carve out farms for themselves there. This was not a colony located by the Hebrew Aid Society, the Jewish Alliance, or the trustees of the Baron de Hirsch Fund, or any similar society or organization; it was formed of Russian Jewish emigrants, who came to make homes for themselves in the United States.

This settlement was established in 1883, the year following the advent of the colony at Alliance; it was comprised of one hundred families, numbering in all about three hundred men, women and children; they selected land which lies partly in Millville and partly in Deerfield township. Each family secured about twenty acres and went to work diligently to clear off the timber and get the land ready for cultivation; while this was being done the women and children were employed in such work as could be obtained to earn enough, with the addition of sums realized from the sales of the wood as it was cut off, to provide food for the family and make the payments for the land as they became due. Rude houses were built by the aid of money secured from the Building Association of the city of Bridgeton, to which, of course, mortgages were given covering the entire properties. This people struggled hard, working from earliest dawn until late at night, with the most determined energy, for a period of seven years; it then became evident that, unassisted, the people could not longer sustain themselves, as the Building Associations were foreclosing the mortgages, and property after property went under the hammer, and the poor settlers were completely disheartened. The outlook for them was indeed gloomy and desperate.

At this terrible crisis of affairs a committee was appointed and sent with an earnest appeal to the late Baron de Hirsch, who was then living, for assistance to extricate them from the difficulties that had crowded upon them and to save their small farms.

The Baron was not the man to turn the deaf ear to cries of his countrymen for assistance, and he sent the sum of $5,000, to be loaned to the struggling people in such sums as careful investigation proved to be needed in each individual case, ranging from $50 to $200; these loans were to bear interest at four per cent., and ten per cent. of the principal with accrued interest, which was to be paid every six months. This timely aid marked a turning point in the history of Carmel settlement; it put heart into the people to renew the struggle, and in the decade that followed they have reached a greatly improved and more comfortable condition. Wealth has not come to the colonists, but they have made sure of a rooting on the land and are earning comfortable livings. In the country from which they came, they had but rough experience in agriculture, and that under entirely different methods from those prevailing here. The success, therefore, which has crowned their efforts to adapt themselves to these new conditions is the more surprising and commendable.

The soil at Carmel is very good, resembling the soil at Alliance and Woodbine; it is a light sandy loam, easily worked, responding readily to manure and fertilizers, and is well adapted to raising vegetables, melons, berries, grapes, peaches, pears, etc., but not heavy enough for cereals, hence the people do not attempt to raise these only in small quantities for their own use. The crops of white and sweet potatoes are very abundant and bring large and sure returns; some of the finest melons produced in New Jersey are raised at Carmel, and the berry and grape crops are of a very high standard.

There are now some very excellent farmers at this settlement; hard workers, who have made a careful study of the capabilities of the soil, who have learned how to treat it to produce the best results and who are constantly on the alert for all that will elevate and improve their condition. The farms are remarkably neat and present a very fine, thrifty appearance. Great care has been exercised to thoroughly free the soil from stumps and roots; there is no sour, soggy land, but the soil is thoroughly broken and pulverized, fed at proper seasons with carefully selected fertilizers and every acre under cultivation yields in rich abundance.

The houses are small and unpretentious, but cozy, comfortable and well furnished.

Like their compatriots at the other settlements, they have fine cows and abundance of excellent poultry. Said one of these settlers, "We do not have to buy much, a little flour and a few groceries; we get our vegetables and fruits from the soil, our milk and butter from the cow, and chickens and the eggs are very good eating. We have enough of these to trade for flour and groceries and from our crops to make the regular payments on the farm and get the plain clothing we need, and, sir, we can afford a holiday suit too." The vineyards, now well set with clusters of Concord grapes, are of very strong and thrifty appearance and the yield is enormous. This colony, which has given special attention to grape culture and the making of wines, is meeting with marked success and finds a rapidly increasing demand for the still wines they manufacture; these wines are of fine body and rich flavor and are rapidly gaining favor among the judges of good wines. The shipping station is Rosenhayn, on the lines of the New Jersey Southern Railroad, and a very busy scene is presented here when the farmers of Carmel and Rosenhayn are shipping their vegetables, fruits and melons to market.

The farm of Isaac Rosen on the extreme southern verge of the Carmel tract, and fronting on the turnpike and the line of the Bridgeton and Millville Traction Company, contains 340 acres, nearly all cleared and in a very high state of cultivation. Mr. Rosen had large experience in farming and ranch life in Texas before coming to Carmel, and made money enough to enable him to secure this large farm. Unlike his fellows at Carmel, he does not grow berries and fruits, but has turned his attention to cereals and potatoes, the soil being somewhat heavier and richer than on most of the farms in the colony; he finds ready market for all he can raise in the cities of Millville and Bridgeton, his farm being about midway between them; he also gives attention to dairying, and has a herd of twenty-two fine cows.

The town site of Carmel is very small, and the synagogue is the only public building. There are three factories, one in which clothing for men and boys is manufactured, and two devoted to the manufacture of ladies' wrappers and waists. In the former, sixty hands are employed, whose wages average from $8 to $10 per week; the other factories employ each twenty-five hands, mostly girls, the average wages being $6 to $7 per week—wages are always paid weekly and in cash. In the establishment of manufactories at this place the Baron de Hirsch trustees have assisted the people. They own the clothing factory and lease the buildings in which the wrappers and waists are manufactured; they also own the machinery in all the buildings.

In contradiction to other colonies, it may be stated that Carmel has had but little extraneous aid, but has reached its present fairly prosperous condition by the indomitable pluck, energy and economy of its people.

OTHER COLONIES.

Attempts have been made to establish Jewish colonies in other localities, but they have all been failures. Of these may be named Mizpah, Atlantic county, six miles from Mays Landing; Reega, eleven miles distant from Mizpah. Malaga, in Franklin township, Gloucester county; Ziontown, four miles distant from the last named; Alberton, near Manumuskin, on the West Jersey and Seashore Railroad; Hebron, on the New Jersey Southern Railroad, near Newfield.

The cause of failure in these attempts at colonization of Jewish families was that they were started by speculators, men whose records did not promise a high order of things for the refugees, and as their capital was very limited, it did not take long to reach the end. The idea of these speculators was that they could get these poor exiles into their clutches, pay them starvation wages and make a big profit out of it, but the Jews failed to be impressed by the glittering inducements held out and did not come in the large numbers anticipated to locate.

Mizpah was projected by a New York firm of cloakmakers, and at one time had one factory, thirteen houses, and probably 25 or 30 Jewish families, numbering in all about 100 persons.

Reega was projected by a wholesale liquor dealer and a picture frame merchant of Philadelphia. A small sewing shop, several frame houses and a small grocery store, with one Jewish family, two Italian and four Polish tried the settlement, but the outlook was so unpromising that they soon fled.

The Jewish families at Malaga found employment in Richman Stocking Factory, and no permanent settlement resulted. Ziontown was projected by a New York coatmaker in conjunction with a Philadelphian, who purchased 1,137 acres of bushland, run it off in town lots at $75 per lot, put up a small sewing shop, and had at one time some seventy persons employed, but work failed and the people were reduced almost to starvation.

The location was all right being somewhat elevated, and if it had been started by the right persons backed by sufficient capital, might have proved successful.

The most notable failure at Jewish colonization was that at Alberton, or, as it is most generally spoken of, Halberton. A New York ticket broker and his nephew, associated with a local speculator under the high sounding title of "The Cumberland Land and Improvement Company," opened up a tract of bushland below Manumuskin Station, on the line of the West Jersey and Seashore Railroad, eight miles south of Millville. A factory, large boarding house, and twelve or fifteen houses were built on the farming tract of a few acres each. The settlers were never in excess of 75, and work soon failed, and they were compelled to leave. There was not sufficient capital back of the concern and it went into the hands of the Sheriff, where it remains, as that functionary has not been able to dispose of it, and the deserted factory and houses, minus doors and windows, present a most forlorn picture to the passengers on the trains that pass the scene of this abortive attempt to found a colony.

Hebron was started by a colonist from Alliance, but never resulted in securing any settlers; it is located in the triangle formed by the railroads at Newfield.

Thus is presented the salient feature of all these South Jersey Jewish colonies. Those established under the auspices of the Hebrew Benevolent Societies are proving successful while every attempt of speculators of start colonies has resulted in complete failure.

Woodbine Public School.

View of Section of Woodbine Containing Workingmen's Houses.

Forcing House, Agricultural School.

Apiary Department of the School.

General View of Woodbine. Industrial Part of Settlement.

Dairy and Cold Storage, Agricultural School.

Dairy, Barn, Silo and Herd at the Agricultural School.

One of the Poultry Yards at the Agricultural School.

New Building of the Baron de Hirsch Agricultural School.

Group of Pupils of Agricultural School, Dairy Department.

Pupils of Woodbine Agricultural School Ready for Milking.

Onion Field on the Agricultural School.

Woodbine Machine and Tool Company.

Synagogue at Woodbine.

Company of Agricultural School Boys with Band.

FURTHER READING.

Bayuk Rappoport Purmell, Bluma and Felice Lewis Rovner. *A Farmer's Daughter: Bluma* (Los Angeles, CA: Hayvenhurst Publishers, 1981).

Brandes, Joseph. *Immigrants to Freedom: Jewish Communities in Rural New Jersey Since 1882* (Philadelphia: University of Pennsylvania Press, 1971).

Eisenberg, Ellen. *Jewish Agricultural Colonies in New Jersey 1882–1920* (Syracuse, NY: Syracuse University Press, 1995).

Goldstein, Philip Reuben. *Social Aspects of the Jewish Colonies of South Jersey* (Philadelphia: University of Pennsylvania Ph.D. thesis, 1921).

Herscher, Uri D. *Jewish Agricultural Utopias in America, 1880–1910* (Detroit: Wayne State University Press, 1981).

Joseph, Samuel. *History of the Baron De Hirsch Fund* ([Philadelphia]: Printed for Baron De Hirsch Fund by the Jewish Publication Society, 1935).

Klein, Moses, et al. *Migdal Zophim & Farming in the Jewish Colonies of South Jersey* (Galloway, NJ: South Jersey Culture & History Center, 2019).

Meyers, Allen. *Southern New Jersey Synagogues: A Social History Highlighted by Stories of Jewish Life from the 1880s–1980s* (Sewell, NJ: Allen Meyers, 1990).

Sabsovich, Katharine. *Adventures in Idealism: A Personal Record of the Life of Professor Sabsovich* (New York: 1922; forthcoming, SJCHC).

Brotman, Richard. *First Chapter in a New Book: a Documentary Portrait of Brotmanville and the Alliance Colony.* DVD documentary (New York: Richard Brotman, 1982).

WHYY. *It's a Mitzvah!: Jewish Life in the Delaware Valley.* DVD documentary (Philadelphia: WHYY, narrated by Larry Kane, 2002).

Yoval. A Symposium upon the First Fifty Years of the Jewish Farming Colonies of Alliance, Norma and Brotmanville, New Jersey (Philadelphia: Westbrook Publishing Co., 1932; republished 1982).

COLOPHON.

Sviatlana Buslovich and Katie Cushinotto edited this text. Tom Kinsella completed the final editing and supervised publication.

The text is set in 13-point Adobe Garamond Pro; section titles are set in 13-point Baskerville Old Face.

THE ALLIANCE HERITAGE CENTER.

Stockton University and its community partners have collaborated to establish a permanent Center with the mission of preserving and disseminating the history and culture of the Alliance Colony and related Jewish farming communities in southern New Jersey. This republication is one in a series highlighting the fascinating history of these communities.

stockton.edu/alliance-heritage

Printed in the USA
CPSIA information can be obtained
at www.ICGtesting.com
CBHW060758271223
2901CB00050B/382